AMAZING MILITARY MACHINES

TANKS

I0812556

BY ELEANOR BRADSHAW

Enslow PUBLISHING

Please visit our website, www.enslow.com. For a free color catalog of all our high-quality books, call toll free 1-800-398-2504 or fax 1-877-980-4454.

Library of Congress Cataloging-in-Publication Data
Names: Bradshaw, Eleanor, author.
Title: Tanks / Eleanor Bradshaw.
Description: Buffalo, New York : Enslow Publishing, [2025] | Series: Amazing military machines | Includes bibliographical references and index.
Identifiers: LCCN 2024002109 | ISBN 9781978539686 (library binding) | ISBN 9781978539679 (paperback) | ISBN 9781978539693 (ebook)
Subjects: LCSH: Tanks (Military science)–History–Juvenile literature.
Classification: LCC UG446.5 .B68268 2025 | DDC 623.74/75209–dc23/eng/20240206
LC record available at https://lccn.loc.gov/2024002109

Published in 2025 by
Enslow Publishing
2544 Clinton Street
Buffalo, NY 14224

Copyright © 2025 Enslow Publishing

Portions of this work were originally authored by Ryan Nagelhout and published as *Tanks*. All new material in this edition is authored by Eleanor Bradshaw.

Designer: Andrea Davison-Bartolotta
Editor: Rachael Morlock

Photo credits: Cover Karlis Dambrans/Shutterstock.com; series art (metal background texture) caesart/Shutterstock.com; series art (title font) TortugaStudios/Shutterstock.com; p. 5 U.S. Marine Corps photo by Cpl. Shannon Kroening/DVIDS; p. 6 U.S. Army photo by Sgt. Charles Probst/Released/DVIDS; p. 7 StockPhotosLV/Shutterstock.com; p. 9 Andrew Skudder/Flickr; p. 10 File:Bundesarchiv_Bild_101I-382-0248-33A,_Im_Westen,_Panzer_II_und_Panzer_I.jpg 1/4 File:Bundesarchiv Bild 101I-382-0248-33A, Im Westen, Panzer II und Panzer I.jpg/Wikimedia Commons; p. 11 Marko Rupena/Shutterstock.com; p. 12 File:Shermandriver.jpg/Wikimedia Commons; p. 13 MichaelD/Alamy Stock Photo; p. 14 File:Charkov 1943 (2).jpg/Wikimedia Commons; p. 15 Sergey Kamshylin/Shutterstock.com; p. 16 Chris Alcock/Shutterstock.com; p. 17 File:USMC M-46 Patton Medium Tank in 1952.JPEG/Wikimedia Commons; p. 18 U.S. Army photo by Spc. Claude Nelson/DVIDS; p. 19 Photofex_AUT/Shutterstock.com; p. 21 Photo by Sgt. Nathan Franco, Fort Irwin Operations Group/DVIDS.0

All rights reserved. No part of this book may be reproduced in any form without permission in writing from the publisher, except by a reviewer.

Printed in the United States of America

Some of the images in this book illustrate individuals who are models. The depictions do not imply actual situations or events.

CPSIA compliance information: Batch #CS25ENS: For further information contact Enslow Publishing, at 1-800-398-2504.

Find us on

CONTENTS

Boldface words appear in Words to Know.

TANK ATTACK

Militaries all over the world use tanks to keep their troops safe and **attack** enemies. Tanks have powerful guns and thick **armor**. They move using tracks that go around a set of wheels. These tracks carry the tank over rough ground.

The tracks on a tank help balance its heavy weight.

IN THE TURRET

The top part of a tank is called a turret. It can spin! Soldiers in the turret can look in any direction while the tank moves forward or backward. The big gun on a tank can fire in any direction.

The large gun on a tank is called a cannon. Most tanks also fire machine guns.

FIRST TANKS

The first tank was named "Little Willie." The British military made it in 1915 to drive over **trenches** during World War I (1914–1918). Little Willie had room for a two-man crew and four gunners. It weighed 18 tons (16.3 mt).

Little Willie was slow and didn't move very well, so later tanks improved its **design**.

WORLD WAR II

By World War II (1939–1945), most countries had tanks in their army. One of the most feared tanks was the German Panzer. The Panzer IV moved almost 25 miles (40 km) per hour. Its large 75-mm gun could hit faraway **targets**.

Heavy armor kept the Panzer's crew of five soldiers safe.

The United States began using its first major tank in 1942. The M4 Sherman could travel as far as 150 miles (241.4 km) at up to 29 miles (46.7 km) per hour. It had a large 75-mm gun on top.

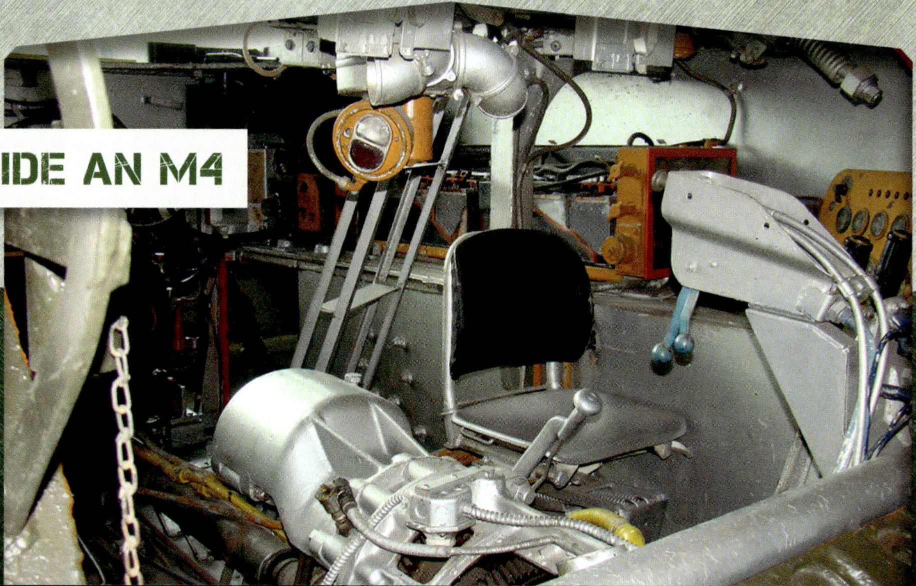

INSIDE AN M4

Crews of five men **operated** the M4 Sherman tanks during World War II.

FASTER AND FARTHER

The Soviet Union's T-34 was another important World War II tank. First made in 1940, the T-34 moved 34 miles (54.7 km) per hour and could travel 268 miles (431.3 km). A crew of four operated the 26-ton (23.6-mt) tank.

The T-34 could travel faster and farther than other World War II tanks.

BATTLE TANKS

After World War II, the main goal of battle tanks was to destroy other tanks. The United States made the M46 Patton in 1950 for this type of **combat**. The U.S. Army used it in the Korean War (1950–1953).

The M46 Patton had a more powerful **engine** for greater speed.

THE M1 ABRAMS

The M1 Abrams is the main battle tank of the U.S. Army. It was first used in 1980. Its heavy gun can blast through thick armor. The tank weighs over 60 tons (54.4 mt) and has a crew of four.

CANNON
TURRET
TRACKS
ROAD WHEELS

INTO THE FUTURE

Tanks of the future will be smaller, lighter, and faster than today's tanks. As war changes, so will tanks. They will have stronger armor to guard their crew, advanced weapons to attack enemies, and more uses in battle.

Speed is important for tanks. They must be quick to attack and withdraw.

WORDS TO KNOW

armor: A thick covering used to keep someone or something safe from harm.

attack: To set upon with force and the intent to harm.

combat: Armed fighting between opposing forces.

design: The pattern or shape of something.

engine: A machine that makes power.

operate: To use and control.

target: The focus of an attack.

trench: A long, narrow cut in the ground used by soldiers in war.

FOR MORE INFORMATION

BOOKS

Brody, Walt. *How Tanks Work.* Minneapolis, Minnesota: Lerner Publications, 2020.

Schuh, Mari. *Tanks and Other Military Vehicles.* North Mankato, Minnesota: Pebble, 2023.

WEBSITES

Britannica Kids: Tank
kids.britannica.com/students/article/tank/277263
This encyclopedia article on tanks focuses on their use in World War I and World War II.

M1A2 Abrams Main Battle Tank
www.military.com/equipment/m1a2-abrams-main-battle-tank
Find specific information about the U.S. Army's number one tank on this website.

Publisher's note to educators and parents: Our editors have carefully reviewed these websites to ensure that they are suitable for students. Many websites change frequently, however, and we cannot guarantee that a site's future contents will continue to meet our high standards of quality and educational value. Be advised that students should be closely supervised whenever they access the internet.

INDEX